AURORA BOREALIS

A Photo Memory

Published by 𝕿𝖔𝖉𝖉 𝕮𝖔𝖒𝖒𝖚𝖓𝖎𝖈𝖆𝖙𝖎𝖔𝖓𝖘
203 W. 15th Ave., Suite 102
Anchorage, Alaska 99501-5128
Phone: (907) 274-8633 • Fax: (907) 276-6858
e-mail: info@toddcom.com

Editor: Flip Todd
Designer: Tina Wallace
Text & Captions: Janice Berry
Photography: Dennis C. Anderson   Hugh Rose
                      Larry Anderson   Nori Sakamoto
                      Wayne Johnson   Todd Salat

Printed in Korea by Samhwa Printing Co., Ltd.
20 19 18 17 16 15 14 13 12 11 10 9

First Edition
ISBN 1-57833-103-X
Library of Congress Catalog Card Number 99-67357

Cover - Photo by Nori Sakamoto

A band of auroras circles through the sky at MacKay Lake, Northwest Territories, Canada in March 1994. The igloo was constructed in half an hour by Inuit (Eskimo) hunters.

Pages 2-3 - Photo by Larry Anderson

For several hours after midnight the Brooks Range was lit by auroras near Dietrich River along the Dalton Highway, 230 miles north of Fairbanks, Alaska in September 1997.

This book is dedicated to Perry and Barbara Schneider, our friends in Fairbanks, Alaska and Don and Penney Sippel, our friends in Whitehorse, Yukon Territory.

AURORA
BOREALIS

A Photo Memory

Left - An auroral corona rises straight overhead above the Kenai Mountains near Seward, Alaska on Prince William Sound.

Right - This spectacular auroral show continued for hours near Homer, Alaska on Kachemak Bay in November 1998.

Below - Vertical rays spawn from an intense burst of light near Haines Junction, Yukon Territory, Canada looking southeast from the Alaska Highway at 1 a.m. on April 17, 1999. A red colored Mars appears low on the horizon just above the trees at right in the constellation Virgo at a time when Mars came closer to the earth than it had in several years.

Photo © Dennis C. Anderson

Photo © Dennis C. Anderson

Spring brings showers in more forms than water as the northern lights rain over the Kenai Mountains. Observers on earth are able to see a partial view of the phenomenon, but astronauts in space can view the entire aurora from the space shuttle.

Right - The red hue in this auroral display appeared at predawn near Anchor Point on the Kenai Peninsula.

Below - A brilliant curtain ripples across the Kenai Range one autumn evening a short distance north of Seward, Alaska.

Next page - This August 22, 1998 display of lights occurred at the start of an active auroral season and shows the constellation Ursa Major or Great Bear, also known as the Big Dipper, over a century-old Russian Orthodox church at the former Russian settlement of Ninilchik on the Kenai Peninsula.
Photo by Dennis C. Anderson

Photo © Dennis C. Anderson

Photo © Todd Salat

Above - *The Hale-Bopp comet burns through space on April 11, 1997 as a subtle band of colorful auroras crosses the horizon outside Homer, Alaska.*

Right - *Thin red bands often precede the start of a good auroral display, sometimes hanging in the sky for several minutes. Looking to the west northwest, this exposure took close to two minutes because the red aurora was so subtle to the human eye. This aurora appeared over Homer during mid-March 1999.*

Next page - *Homer residents were treated to an early show of crowning, surging auroras at 8 p.m. one evening in November 1998, an unusually early hour for such a good display.*

Left - This ultra-bright corona (looking straight up into the base of an aurora) occurred at 10 p.m. directly over Anchorage. It was easily seen despite the street and house lights.

Below - The auroras out-shone the urban lights of Anchorage for over three hours in February 1995.

Right - The red afterglow from a sunset that disappeared to the human eye an hour earlier over Earthquake Park in Anchorage, appears on film due to the long exposure. This scene with Mount Susitna in the background occurred on April 12, 1990 at about 11 p.m.

Photo © Larry Anderson

Photo © Wayne Johnson

13

Right - Anchorage lights twinkle below a three-quarter moon and a subtle stream of pale green auroras, the most common auroral color, one April evening.

Below - Triple bands low on the horizon are reflected in Mirror Lake near Eagle River, Alaska with Bear Mountain in the background.

Next page - Cloud-like blue auroras surround a rare rose-colored auroral veil, pouring light over Hatcher Pass in the Talkeetna Mountains north of Anchorage in February 1994.

Photo © Wayne Johnson

Photo © Wayne Johnson

Left - This auroral band bolted flame-like into the night sky above the photographer's Hatcher Pass cabin in January 1994.

Right - Twin vertical bands enhance a full moon over Mount Susitna, 65 miles west of Anchorage, with Cook Inlet in the foreground.

Below - Auroral waves sweep the sky above the Matanuska Glacier, with the Talkeetna Mountains to the north.

Photo © Wayne Johnson

Photo © Wayne Johnson

Left - Even scientists, who have become more accurate at predicting auroras according to the 11-year solar sunspot cycle, were surprised by the appearance of these January lights over Pioneer Peak in the Chugach Mountains near Palmer, Alaska, 40 miles north of Anchorage.

Below - A green corona beams into the darkness above Hatcher Pass, while the Big Dipper appears inverted under the duck's head.

Right - Tall purple rays fill a partly cloudy sky in February near sunrise at Hatcher Pass.
Photo by Todd Salat

Photo © Wayne Johnson

Photo © Todd Salat

Right - A burst of red and white auroras mingles with green in an auroral corona near Hatcher Pass above spruce trees. The Big Dipper is barely visible near the zenith, or apex, of the corona.

Below - Most auroras are green, the result of particles streaming into the atmosphere at speeds of up to 37,000 miles a second. The multi-colored effect of red and green over spruce trees in Hatcher Pass that occurred in February 1994 is caused by a mixture of slow and fast-moving particles.

Photo © Todd Salat

Photo © *Todd Salat*

Left - *A multi-colored aurora appears over the Alaska Range and Black Rapids Glacier beneath a full moon. Orion's belt appears above the moon in this scene captured at about 4 a.m. in November 1998.*

Right - *Auroral movements in the night sky often help break the monotony of long trips by northern truck drivers such as the one driving this 18-wheeler just south of Fairbanks on the Parks Highway.*

Below - *A wide-banded aurora curls itself across the night sky above spruce trees in September south of Cantwell and east of Denali National Park.*

Next page - *A broad band extends across Friday Creek Valley near the shore of the Knik River in the Chugach Mountains of Alaska.* **Photo by Wayne Johnson**

Photo © Wayne Johnson

Photo © Wayne Johnson

Photo © Wayne Johnson

Left - *A flash of green auroras appears from behind the Talkeetna Mountains. The deserted remains of Independence gold mine are in the foreground on a -20°F (-29°C) January evening.*

Right - *Tall auroral rays create a transparent wall of light over a near full moon seen along Alaska's Richardson Highway in the Alaska Range in November 1998. The blue-purple cast typically results from sunlight intersecting the high atmosphere above the observer in a process called "resonance scattering."*

Below - *A pair of auroral arcs curve and twist across the sky in a view from the Hurricane Gulch Bridge at Mile 174 of the Parks Highway between Fairbanks and Anchorage. The temperature was -20°F (-29°C) at 4 a.m. March 1998.*

Photo © Todd Salat

Photo © Todd Salat

Photo © Todd Salat

Left - *Multiple auroral bands cover the southern sky looking down the Richardson Highway along the Alaska Range in November 1998.*

Below - *A blood-red patch of auroras seems drawn by the green "magnet" just above the Alaska Range along the Richardson Highway. The 48-inch diameter trans-Alaska crude oil pipeline runs along the horizon at center.*

Right - *Long ribbons of green auroras trail across the southwest sky, covering a three-quarter moon above the Flattop Mountain trailhead south of Anchorage about 11 p.m. in early April. A small cluster of lights from Anchorage homes twinkle at the front of the mountains.*
Photo by Wayne Johnson

Photo © Todd Salat

Left - Cloud-like auroras fill the horizon behind the Wolf Park Lodge at Mile 156 of the Parks Highway north of Trapper Creek in March.

Below - A scythe-like green aurora dances above moonlit Wolverine Peak east of Anchorage in the Chugach Mountains in April.

Right - Moonlight illuminates 20,320-foot Mt. McKinley as a powerful wave of red and green auroral curtains surge overhead in March 1998.

Photo © Wayne Johnson

Photo © Wayne Johnson

31

Right - *A giant eddy of light loops its way across the Big Dipper in the night sky. The aurora hangs over Rainbow Ridge in the Alaska Range north of Paxson on the Richardson Highway in November 1998.*

Below - *The headlights of a large truck zoom across the Hurricane Gulch Bridge below nature's headlights at Mile 174 (174 miles from Anchorage) over the deepest gorge between Fairbanks and Anchorage on the George Parks Highway.*

Next page - *Shimmering curtains of green auroras are reflected by a tundra pond in Denali National Park and Preserve.* **Photo by Hugh Rose**

Right - Auroras swirl above Wolverine Peak, at right, beneath a three-quarter moon in April near the Flattop Mountain trailhead south of Anchorage.

Below - A series of auroral arcs emerge above Mirror Lake north of Anchorage in October. The auroras were active for five days in a row in October, 1997.

Next page - A light-colored auroral rayed band resembling an arch terminates in a rose auroral patch over Byers Lake, Denali State Park on September 10, 1992.
Photo by Larry Anderson

Photo © Wayne Johnson

Photo © Wayne Johnson

*A broad display of vertical rays hangs above
Mount McKinley as seen from a remote spot
along the Petersville Road south of North
America's highest mountain.*

Three auroral bands erupt from behind 16,390 foot Mt. Blackburn (4,996 meters) in the Wrangell Mountains east of Glennallen, Alaska in October 1998. Mt. Blackburn was better lit than 6,696 foot Mt. Donohoe (2,040 meters) at right, by a brilliant full moon.

Photo © Hugh Rose

Above - *Green and yellow auroral bands flare over the Endicott Mountains high above the Arctic Circle in the Brooks Range.*

Left - *Bright green bands rise into the sky over the Endicott Mountains in the Brooks Range above the Dalton Highway leading to Prudhoe Bay, Alaska on September 5, 1997. The display lasted nearly three hours and was the brightest the photographer had seen in three years.*
Photo by Larry Anderson

Right - *An emerald green aurora forms a J in the sky over the foothills of the Brooks Range on the North Slope.*

Next page - *Multiple arcs and bands materialized out of thin air to perform an all-sky show over the Wrangell Mountains above the historic Kennecott Mine in October 1998. During the 12-second exposure the mine was lit with a spotlight. The mine, rich in copper deposits, operated from 1910 to 1938 with the backing of Guggenheim and J.P. Morgan financial interests.*
Photo by Todd Salat

Photo © Todd Salat

39

Above - *A curtain of green auroras glows over the Philip Smith Mountains which travel southwest to northeast from Atigun Pass to the Romanzof Mountains within the Brooks Range. Drained to the north by the Sagavanirktok River which flows into the Beaufort Sea near Prudhoe Bay, many of Alaska's most productive oil-bearing formations are exposed in its stream valleys.*

Below - *Green and red auroras rise over the Endicott Mountains in the Gates of the Arctic National Park and Preserve in the central Brooks Range.*

Right - Translucent streamers flicker above the 800-mile trans-Alaska oil pipeline near Mile 243 of the Richardson Highway on November 7, 1998. The pipeline is elevated in regions of permafrost. Magnetic storms associated with the aurora induce currents of close to 1,000 amperes in the pipeline causing corrosion in its half-inch thick walls.

Below - Neon green bands swirl over the Pool Arctic No. 6 exploratory oil rig near Umiat in the Brooks Range foothills on the North Slope of Alaska at -40°F (-40°C).

Next page - Pale blue beams of light shoot skyward north of the Arctic Circle at about 68° North latitude in the Philip Smith Mountains. This part of Alaska's Brooks Range is in the Arctic National Wildlife Refuge adjacent to Canada's Yukon Territory. **Photo by Hugh Rose**

Photo © Todd Salat

Photo © Todd Salat

Left - Green ribbons of northern lights appear to emanate from the chimney of a cabin on the Stampede Trail near Healy, Alaska. Solar winds comprised of protons and electrons take two to three days to reach the earth from the sun and cause auroras simultaneously in the northern and southern hemispheres.

Right - As the moon begins to rise above the horizon, curving auroral bands sweep above the town of Circle, population 86, on the banks of the Yukon River in January 1999. The temperature was -55°F (-48°C), so cold the photographer's film became brittle and broke in two as it was advanced.

Below - A circular aurora appears over the town of Circle, Alaska, a short distance south of the Arctic Circle (66° 33 minutes North latitude). Looking northeast from Eagle Summit on the Steese Highway, this photo was taken beneath a full moon in -40°F (-40°C) weather.

Photo © Todd Salat

Photo © Todd Salat

Right - An emerald arc trails hundreds of miles across a December sky at midnight, behind a backlit white spruce tree on Murphy Dome west of Fairbanks. The Big Dipper seems to rest atop the tree in an area with some of the best aurora viewing in North America.

Below - A palette of green, blue and purple auroras paints the horizon over the Philip Smith Mountains in the Arctic National Wildlife Refuge. Auroras are like neon lights; gases in the upper atmosphere, nearly a vacuum, turn colors as they are electronically activated.

Next page - This purple aurora with a light fringe toward the bottom appeared to the south of Muncho Lake, British Columbia on the Alaska Highway on the evening of May 4, 1999.

Photo © Todd Salat

Photo © Hugh Rose

Right - A tornado-like display of auroral bands dominates the sky over Vee Lake in Yellowknife, Northwest Territories, Canada in September 1993. The auroras appear to be touching the horizon, occurring 1,000 miles (1,609 kilometers) away from the observation point and 65 miles (105 kilometers) above earth.

Below - The Hale-Bopp comet casts a twin tail, one blue and one white, in the night sky above Wulsh Lake near Yellowknife, the capital of Canada's Northwest Territories. This image was taken in a five-minute time exposure which made the northern lights, barely visible to the human eye, far stronger in intensity than they would have been otherwise. The photographer used a star tracking device, which rotated during the exposure to keep the stars from appearing as arcs, as they normally do in time exposures. The white comet tail points away from the sun and is comprised of water, carbon dioxide and organic matter. The blue tail is comprised of ionized gases emanating from the solar wind. Unlike sunlight that is emitted straight from the center of the solar system, solar winds are emitted in a swirl, like a rotating lawn sprinkler.

Left - *A cloud of auroras swirls above Vee Lake in the Northwest Territories and combines with moonlight to light a snow-covered cliff in April 1999.*

Right - *A plume of vertical rays bursts into the northern dusk over Great Slave Lake near Hay River in the Northwest Territories, Canada. This display occurred looking northwest two hours after sunset at midnight in early May 1994. This is the best season for photographing the aurora because of richer image colors and good weather.*

Photo © Nori Sakamoto

Previous page - Cool, flame-like auroras shoot from the eastern horizon overlooking frozen Cameron Falls near Yellowknife, Northwest Territories, Canada in late March 1994.

Left - An aurora spreads rays in almost every direction above Cameron Falls near Yellowknife, Northwest Territories, Canada in late March 1994. The full moon lit up the snowy valley almost as bright as daylight.

Below - The Hale-Bopp comet sails through an early evening display of purple and green auroras one and one-half hours after sunset over frozen Vee Lake near Yellowknife, Northwest Territories in April 1997.

Next page - After a quiet night of low activity for the northern lights, a two-toned display of auroras began to blaze above the northwest horizon from the shores of Great Slave Lake near Yellowknife, Northwest Territories in early April 1995.
Photo by Nori Sakamoto

Photo © Nori Sakamoto

Photo © Nori Sakamoto

Right - The moon rising in the eastern sky imitates a sunset while auroral rays pulse over frozen Great Slave Lake, Northwest Territories, Canada about three hours after sunset in April 1998.

Below - An auroral arc forms a half-circle in the northern sky over frozen Cameron Falls near Yellowknife, Northwest Territories, on April 1, 1997.

Right - Moonlit birch trees reach skyward toward the dancing aurora at the Fort Providence, Northwest Territories campground in early May 1999.

Below - The moon's reflection appears to the west on the edge of frozen Prelude Lake in April 1999 near Yellowknife, Northwest Territories. Yellowknife is located at about 62°N latitude making it prime aurora-viewing territory for its longitude.

Photo © Nori Sakamoto

61

Left - *Auroras radiate in the western sky at Great Slave Lake near Yellowknife, Northwest Territories in mid-April 1998 with the photographer's tent in the foreground. When the photographer retreated to his tent in the -30°F (-34°C) cold and boiled water, the steam formed a small, cloud and shortly thereafter it began to snow inside the tent!*

Right - *Bright auroras unfold over the Ogilivie Mountains from along the Yukon Territory's Dempster Highway beginning in late August 1998.*

Below - *A magnetic storm results in auroras filling the southern horizon near Great Slave Lake on a road about 15 miles from Yellowknife. The mirror-like surface of the lake which was blown clear of snow reflects the auroras in early April 1994.*

Next page - *Brilliant moonlight, which makes the aurora more difficult to see, illuminates the clouds above Prelude Lake, Northwest Territories in April 1999.*

Photo © Nori Sakamoto

Photo © Nori Sakamoto